S0-BYW-823

MOTHER NATURE IS TRYING TO

SAVAGE ANIMALS

BY JANEY LEVY

Gareth Stevens
PUBLISHING

Please visit our website, www.garethstevens.com. For a free color catalog of all our high-quality books, call toll free 1-800-542-2595 or fax 1-877-542-2596.

Library of Congress Cataloging-in-Publication Data

Names: Levy, Janey, author.
Title: Savage animals / Janey Levy.
Description: New York : Gareth Stevens Publishing, [2020] | Series: Mother nature is trying to kill me! | Includes index.
Identifiers: LCCN 2018055314| ISBN 9781538239780 (paperback) | ISBN 9781538239803 (library bound) | ISBN 9781538239797 (6 pack)
Subjects: LCSH: Dangerous animals–Juvenile literature.
Classification: LCC QL100 .L495 2020 | DDC 591.6/5–dc23
LC record available at https://lccn.loc.gov/2018055314

First Edition

Published in 2020 by
Gareth Stevens Publishing
111 East 14th Street, Suite 349
New York, NY 10003

Copyright © 2020 Gareth Stevens Publishing

Designer: Sarah Liddell
Editor: Monika Davies

Photo credits: Cover, p. 1 Audrey Snider-Bell/Shutterstock.com; background used throughout Lana Veshta/Shutterstock.com; p. 4 Willyam Bradberry/Shutterstock.com; p. 5 sonya etchison/Shutterstock.com; p. 7 walter stein/Shutterstock.com; p. 9 (main) Anton_Ivanov/ Shutterstock.com; p. 9 (inset) JT Platt/Shutterstock.com; p. 11 Henk Bogaard/ Shutterstock.com; p. 13 Paul Tessier/Shutterstock.com; p. 15 Clement Carbillet/Biosphoto/ Getty Images; p. 17 (main) sevenke/Shutterstock.com; p. 17 (inset) Fazwick/ Shutterstock.com; p. 19 davemhuntphotography/Shutterstock.com; p. 20 Michal Ninger/ Shutterstock.com; p. 21 (wild pig) krslynskahal/Shutterstock.com; p. 21 (African buffalo) EcoPrint/Shutterstock.com; p. 21 (hyena) Vladimir Wrangel/Shutterstock.com; p. 21 (bison and background) Betty Shelton/Shutterstock.com.

All rights reserved. No part of this book may be reproduced in any form without permission in writing from the publisher, except by a reviewer.

Printed in the United States of America

CPSIA compliance information: Batch #CS19GS: For further information contact Gareth Stevens, New York, New York at 1-800-542-2595.

CONTENTS

Words in the glossary appear in **bold** type
the first time they are used in the text.

BARBAROUS BEASTS

Perhaps, like many people, you have a pet dog or cat. Or, maybe you enjoy feeding wild ducks. These can all be wonderful **experiences** with the animal world. But be careful—some animal experiences can be deadly!

When you think of deadly animals, you might first think of large predators, such as wolves and sharks.

GREAT WHITE SHARK

However, these animals don't kill or hurt people very often. The most dangerous and deadliest animals include plant-eating **mammals**, some **reptiles,** and a giant bird!

IT'S IMPORTANT TO REMEMBER THAT ALL ANIMALS, EVEN THE CUTE ONES, MAY BE DANGEROUS. WE SHOULD TREAT ALL ANIMALS WITH KINDNESS AND RESPECT

EVIL ELEPHANTS

Elephants are popular and friendly characters in many stories. You may have even seen one of these smart, huge **herbivores** in real life. But don't be fooled by how the storybooks describe elephants. These are extremely dangerous animals!

Elephants are the largest land animals alive today. Even though they're

THE FORCE OF NATURE

ELEPHANTS CAN WEIGH UP TO 16,500 POUNDS (7,484 KG) AND GROW UP TO 13 FEET (4 M) TALL. THESE HUGE ANIMALS CAN RUN UP TO 25 MILES (40 KM) PER HOUR!

quite large, elephants can run fast—much faster than most people can. They may charge unexpectedly and for no apparent reason. They might even use their great weight to **trample** you to death!

WHEN AN ELEPHANT HOLDS ITS EARS OUT TO THE SIDE, IT'S PRETENDING TO CHARGE. IF AN ELEPHANT HOLDS ITS EARS BACK FLAT AGAINST ITS HEAD, THOUGH, WATCH OUT!

HORRIBLE HIPPOS

Hippos are huge mammals found in Africa. They're herbivores and often considered cute and gentle creatures. But don't be fooled. Hippos are always ready to attack. They're one of the deadliest animals in the world!

Hippos have a giant mouth that opens wide. At the front of their mouth are long, sharp teeth used for fighting. A hippo's **canine** teeth can grow up to 20 inches (51 cm) long. On land, hippos can run over 15 miles (24 km) per hour. You can't escape an

THE FORCE OF NATURE

HIPPOS CAN GROW UP TO 16.5 FEET (5 M) LONG AND WEIGH OVER 9,900 POUNDS (4,490 KG).

BRUTAL BLACK RHINOS

Like hippos, black rhinos are huge mammals that eat plants and live in Africa. You might think herbivores are harmless, but as hippos prove, that's not always true. Black rhinos are also deadly!

Black rhinos have a good sense of hearing and smell.

THE FORCE OF NATURE

BLACK RHINOS WEIGH UP TO 3,100 POUNDS (1,406 KG). THEY CAN RUN AROUND 40 MILES (64 KM) PER HOUR IN SHORT BURSTS. IMAGINE ALL THAT WEIGHT COMING AT YOU AT FAST SPEED!

However, they can't see very well. They're also usually in a bad mood. If a black rhino hears an unfamiliar sound or smells something unusual, it may suddenly charge. Their weight and sharp horns

BLACK RHINOS HAVE TWO HORNS ON THEIR NOSE. THE FRONT HORN IS THE LARGER OF THE TWO AND CAN GROW UP TO 50 INCHES (127 CM) LONG!

MENACING MOOSE

Moose are large herbivores with sweet, goofy faces. They prefer to avoid people. But moose and people still meet—often by chance—and these meetings can quickly turn dangerous. In Alaska, there's a greater danger that you'll be hurt by a moose than by a bear!

Moose often attack using their powerful front legs and sharp hoofs to stomp and kick their enemies. Female moose may hurt people when **protecting** their calves, or young. Male moose are most dangerous and ready to attack during **mating** season.

THE FORCE OF NATURE

MALE MOOSE CAN BE MASSIVE! THEY CAN GROW UP TO 7 FEET (2.1 M) TALL AND WEIGH OVER 1,300 POUNDS (590 KG).

IF MOOSE THINK THEY'RE IN DANGER, THEY MAY CHASE YOU. THEIR LONG LEGS CAN HELP THEM RUN UP TO 35 MILES (56 KM) PER HOUR! HOWEVER, THEY'LL USUALLY ONLY CHASE YOU A SHORT DISTANCE.

AWFUL OCELLATED CARPET VIPER

Do you fear snakes? Many snakes are helpful and eat pests that bother humans. But others—such as the ocellated carpet viper—are worthy of your fear. These snakes have a bite that can kill.

Ocellated carpet vipers are found in western Africa. They bite with sharp teeth

THE FORCE OF NATURE

THE OCELLATED CARPET VIPER IS OFTEN CALLED THE DEADLIEST SNAKE IN AFRICA. SOME SAY THESE SNAKES MAY KILL AS MANY AS 20,000 PEOPLE EACH YEAR!

called fangs, which carry **venom**. The venom causes terrible pain and swelling around the bite. Then, the venom causes bleeding inside your body, and blood might come out of your mouth, nose, and ears.

THE OCELLATED CARPET VIPER ISN'T A VERY LONG SNAKE.
IT ONLY GROWS TO BE ABOUT 2 FEET (0.6 M) LONG.

15

CRAZED CASSOWARIES

You're probably familiar with birds you've seen in your backyard, such as robins, cardinals, and blue jays. These gentle birds are fun to watch. However, not all birds are harmless.

The cassowary is a large bird that can actually kill you! Cassowaries are too large and heavy to fly, but they can run fast, jump high, and kick hard. They also have sharp talons, or claws, on their feet. They can break your bones and cut you open. Don't mess with a cassowary!

THE FORCE OF NATURE

SOUTHERN CASSOWARIES CAN GROW OVER 5 FEET (1.5 M) TALL AND WEIGH UP TO 128 POUNDS (58 KG).

THE SOUTHERN CASSOWARY HAS BEEN CALLED THE WORLD'S MOST DANGEROUS BIRD. THEY'VE ATTACKED OVER 200 PEOPLE.

TALONS

17

NASTY NILE CROCODILES

There's one animal that most people agree is scary and dangerous: the crocodile. These large, heavy reptiles are **carnivores** with powerful **jaws** filled with sharp teeth. Nile crocodiles kill more people in Africa each year than any other animal.

Nile crocodiles are fierce and scary predators. They

THE FORCE OF NATURE

NILE CROCODILES CAN GROW TO 20 FEET (6 M) LONG. THE AVERAGE NILE CROCODILE WEIGHS AROUND 500 POUNDS (226 KG), BUT SOME CAN WEIGH THREE TIMES MORE!

mostly eat fish. However, when a Nile crocodile is hungry, any animal will do for food—including you! Get too close to the water where they live and you could end up being their next meal.

THE NILE CROCODILE IS ONE OF THE MOST DANGEROUS CROCODILES.
IT'S BELIEVED NILE CROCODILES ATTACK OVER 300 PEOPLE EVERY YEAR.

19

WHY ANIMALS ATTACK

By now it might seem as if wild animals are out to get you, but that's not exactly true. Life is hard for wild animals. They face constant danger from other animals and people. Predators and people hunt them. They battle with their own kind for food, territory, and mates.

Wild animals don't want to attack you, but they will if necessary to protect themselves and their young. Keeping your distance from wild animals is the best way to stay safe.

MORE DANGEROUS ANIMALS

WILD PIG

AFRICAN BUFFALO

HYENA

BISON

GLOSSARY

canine: a long, pointed tooth near the front of the mouth

carnivore: an animal that eats meat

experience: something that you have done or that has happened to you

herbivore: an animal that only eats plants

jaws: the bones that hold the teeth and make up the mouth

mammal: a warm-blooded animal that has a backbone and hair, breathes air, and feeds milk to its young

mate: to come together to make babies. Also, one of two animals that come together to produce babies.

protect: to keep safe

reptile: an animal covered with scales or plates that breathes air, has a backbone, and lays eggs, such as a turtle, snake, lizard, or crocodile

trample: to injure by stepping heavily on something or someone

venom: something an animal makes in its body that can harm other animals

FOR MORE INFORMATION

BOOKS

DK Publishing. *Nature's Deadliest Creatures: Visual Encyclopedia.* New York, NY: DK Publishing, 2018.

Jenkins, Steve. *Deadliest!: 20 Dangerous Animals.* New York, NY: Houghton Mifflin Harcourt, 2017.

Price, Sean Stewart. *The World's Deadliest Animals.* North Mankato, MN: Capstone Press, 2017.

WEBSITES

Black Rhinoceros
kids.nationalgeographic.com/animals/black-rhino
Find out more about the fierce black rhino.

9 of the World's Deadliest Mammals
www.britannica.com/list/9-of-the-worlds-deadliest-mammals
This list of nine of the world's deadliest mammals includes some that are sure to surprise.

Snakes: Facts
idahoptv.org/sciencetrek/topics/snakes/facts.cfm
Learn more about snakes, including what they eat and how they move.

Publisher's note to educators and parents: Our editors have carefully reviewed these websites to enscied.ucar.edu/webweatherwever, and we cannot guarantee that a site's future contents will continue to meet our high standards of quality and educational value. Be advised that students should be closely supervised whenever they access the internet.

INDEX